the social pandemic
the influence and effect of social media on modern life

D1419972

For Katie Blighton
The love of my life

for being you!

about the author

I am 25 years old, I live in Leicestershire, England, the county known as the inspiration for "the shire" in Tolkien's classic, The Lord of the Rings. I am an avid lover of technology, as well as movies and music.

I run my own Social Network dedicated to Dodgeball players called iDodge[1]. This is likely to be the first of many. I have an iPhone game in the app store called pShooter that allows you to shoot your friends and family, it seems unlikely that I will make any more, although I will update it, and produce an android and a Windows Phone version. I do PC Repairs in the area surrounding my home, as well as being a partner in a Social Media marketing company[2]. I am working on a new website for car enthusiasts, it will be a cross between Twitter and Auto-Trader. This will not be another "me too" website but rather a revolutionary product in a number of ways.

You can, of course, get in touch and keep up to date with me via Social Media. On Twitter (@giles_lloyd), or on my personal blog, gileslloyd.blogspot.com.

---

[1] www.i-dodge.com

[2] www.xanadumarketinggroup.com

contents

introduction

"Interdependence is and ought to be as much the ideal of man as self-sufficiency. Man is a social being."
- Mahatma Gandhi

As human beings, we crave social interaction. The need for it is built into our genetic makeup. Loneliness is linked to depression, and people with healthy social lives are known to live longer on average. Over the course of human history, the desire remains constant, however the means by which we fulfil this need is constantly changing and evolving. The most fundamental changes are driven, in the largest part, by technology.

The invention of the telephone was the first major change, allowing us to communicate with each other over long distances in real time for the first time ever. However, telephone's were expensive and therefore only available to the upper classes.

Next came the introduction of the mobile telephone. It wasn't until the mid 90's, when mobile phones became more widely available with handsets to suit every budget, that this technology began to revolutionise our social lives. Now, not only could we communicate over long distances in real-time, we all had our own personal number, including young children. We could be contacted wherever we were, whenever. It was at this time that the first widespread text-based communication revolution happened, the introduction of sms, or text messaging. This was widely adopted by the younger generations and became a multi-million pound business overnight. This set the stage for a much larger revolution that was looming on the horizon.

Around this time, the Internet was still in its infancy. In the early days of the Internet, legal restraints that deemed making money from it a criminal activity, held back its growth. In the mid 90's these sanctions were lifted, and the Internet that we now know and love was born. The commercial Internet attracted entrepreneurs who began to create multi-million pound businesses online. A global network connecting billions of people from around the globe was the perfect breeding ground for a social revolution, and as people the world over began to connect to the web in droves, it was only a matter of time before it happened.

In recent years, with the advent of social networking giants such as Facebook, YouTube and Twitter, Social Media has exploded. However, it is not a new phenomena. As early as the birth of the commercial internet, Social Networking has existed in one form or another. Mass adoption by younger generations, of instant messaging services such as MSN Messenger paved the way for the onslaught of Social Media in it's current form. The earliest Social Media sites were online havens for teens, with no credibility with the older generations, their influence was not nearly as widespread as it is today. Some of these sites still exist, however their market share is dropping rapidly. Bebo, one of the earliest Social Media juggernauts, recently collapsed, MySpace has been bought out several times in recent years as it struggles to find it's place in the new world of social media. Meanwhile Facebook and Twitter continue to grow. At the time of writing, it has been a matter of weeks since Facebook went public, at a $100 Billion valuation. The largest tech IPO in history. Last year its reported net earnings were $1 Billion. A reflection of the somewhat prophetic line from "The Social Network", the hollywood blockbuster about the creation of Facebook,

"A million dollars isn't cool. Do you know what is cool? A

BILLION dollars!"

YOU DON'T GET TO 500 MILLION FRIENDS WITHOUT MAKING A FEW ENEMIES

Columbia

(The Social Network - Movie Poster)

In the application for Facebook's IPO, they made their goal clear. There are currently over 2 Billion internet users, and Facebook wants to connect every single one of them. Their own figures indicate that in many countries they already have over 80% penetration rates.

The fundamental difference between Social Media as it exists today and older services such as Bebo, MySpace and MSN is the wider target audience. While the older sites were mostly populated by teenagers, now older generations are beginning to take Social Media seriously. The fastest growing age group on Facebook at the time of writing is 30+. With the more widespread adoption of these services, the real potential that they offer is becoming apparent. The Social Revolution is in full swing, and what we have seen so far is the tip of the iceberg, we have barely scratched the surface of what is possible through the power of Social Media.

The days where being a technophobe would not affect a person's life one iota are long gone. Social Media has the power to enrich our lives, and relationships on both a personal and professional level.

However, while there are countless positive benefits, there is also the vast potential for misuse and there are many potential negative effects of the Social Media Revolution. In the following chapters we will be exploring both the positive and negative effects of Social Media on every aspect of our lives.

Recently, the UK government announced plans that would ensure every household in the UK would eventually have access to the Internet. The implication being that the Internet is now considered an amenity. Access to the web is now viewed as important as access to clean water, gas and electricity to maintaining quality of life, which, in turn, raises questions about the longevity of net neutrality.

"Net Neutrality" is the perception that everything that resides on the web is of equal value to the consumer. The suggestion being that at some undetermined point in the future, ISP's will collectively abandon the pricing structure that they have today, and adopt one akin to the package subscriptions offered by "cable" or Satellite Television providers. Theoretically, with this type of tiered pricing structure, the consumer would pay a basic, monthly fee for general access to the web. However, if the consumer required access to sites such as Facebook or Twitter, they would have to add a "Social Media Package" to their subscription, at an additional cost. Just as you have to add a "Sports Package" or "Movies Package" to your Television subscription in order to view specific channels. Likewise, if you require access to YouTube or similar sites, you would need to add a "Media Package" to your broadband subscription. And so on. This concept clearly undermines the idea of the internet as an amenity, and has met with severe global backlash.

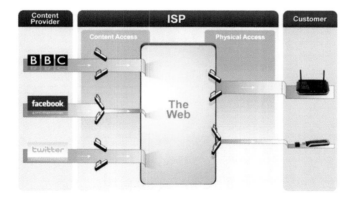

3

---

[3] A visual depiction of ISPs without Net Neutrality. Image source: www.betransformative.com

On May 8th 2012, The Netherlands became the first country, at least within the EU, to pass a law securing net neutrality long-term, forcing ISPs to become "blind" to the traffic they carry, and treat all of it equally. In comparison, net neutrality has not been specifically discussed in the UK since 2006. A debate was held on the subject at Westminster on 20th March 2006. The conclusion was that "Net Neutrality laws in the UK would be extreme, unattractive and impractical." In the US, 5 separate attempts have been made to pass bills seeking to prevent service providers from using variable, or tiered pricing models. All of them have failed.

At least for now though, access to Social Media is easy and free. It is influencing our lives in very real ways and, in all likelihood, will continue to do so for years to come.

friendship

"If a man does not make new acquaintance as he advances through life, he will soon find himself left alone. A man, Sir, should keep his friendship in constant repair."
- Samuel Johnson (1709 - 1784)

The primary goal of Social Media is to connect people, creating and enriching friendships and it is in this category that Social Networks get the most use. While Social Networks exist that are designed to help you meet new people, such as Tagged, the market leaders, primarily Facebook, are designed to be online platforms to keep in touch with your "real world" friends, the people that you already know. Although Facebook can, and does help people expand their social circles, this is through mutual friends, rather than complete strangers.

As Social Media continues to become an integral part of our lives, sites such as friends reunited will have to change direction or risk becoming redundant. More and more people are getting in touch with old friends via their current social networks. And as the younger generations mature and leave school, they will simply never lose touch with their friends, they will always be connected via Social Media.

The primary purpose of Social Networking sites is to allow us to quickly and easily share our lives with the friends and family, who may not otherwise be kept up to date. This is done via status updates as well as media sharing, by posting photo's, videos, and links. Some people however, take this to an extreme, by sharing every minute, insignificant detail of their daily activities. You need only spend 5 minutes on Twitter to realise the vast numbers of people tweeting about feeding their cat and fetching milk from the shop. This kind of mundane chatter dilutes the quality of the content on Social Media sites, and you will find, should you engage in this type of activity, that your real world friends will remove you from their Social Networks in droves.

The definition of "friend" in Social Media differs from the definition we would use in the real world. On Facebook, for example, to contact or share with anybody, the 1st thing you must do is send that person a friend request. Should they accept this request, you will be in each other's "friend" list. Then you may contact or share with that person freely. However, while Facebook makes no distinction, in real life, you may not describe that person as a friend at all. They may be a relative, colleague, teacher, employer or merely an acquaintance. According to Dunbar's number, the average person has 2-3 close friends, with a wider circle of up to 150. In comparison, the average facebook user has between 200 and 400 people in their friends list.

"True happiness consists not in the multitude of friends, but in their worth and choice."
- Samuel Johnston

While a larger friends list makes a person feel more popular, the reality is, that this is not an accurate reflection of their true Social Circle. As any Facebook user will tell you, they would probably not even say hello to at least 10% of the people in their friends list, if they walked past them in the street. In fact most people have at least 5 people in their friends list that they would not even *recognize* if they walked past them in the street. Google has attempted to correct this flawed system with their own social network, Google+. Google+ is the "new kid on the block" of Social Networking, since it was only launched to the general public on September 20th 2011. However, it already has a massive user base of 90 Million members and is expected to reach 400 Million by the end of 2012. In comparison, Facebook launched to the public in 2006 and didn't reach 400 Million users until 2010. Instead of having a single list of "friends" that incorporates all of your contacts, Google+ has "Circles". Which are essentially lists that you define yourself, in order to more accurately reflect the social circles that you have in real life. When you share updates and media on Google+, you can choose to share it with everyone, or just specific circles, such as your close friends or family. Google circles is Google's attempt to fix the "broken" way that we share online. This is the description of Circles from Google+:

"You share different things with different people. But sharing the right stuff with the right people shouldn't be a hassle. Circles make it easy to put your friends from Saturday night in one circle, your parents in another, and your boss in a circle by himself, just like real life."

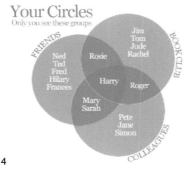

When you post or share content...

Give visibility to specific circles
This means any of the people in those circles who look at your stream will see that content. Other people will not.

If one of the people in those circles also has you in one of their circles, your item will appear in their stream.

Disallow re-sharing
If you make content visible only to specific people you can prevent it being re-shared to a wider network.

---

4 Google Circles Infographic. Image Source: www.dotponto.com

Other Social Networks have noticed this anomaly and sprung into existence to help fill the gap, catering to much more specific relationship types. There is Cupple, a Social Networking app for you and your partner. It's not really "networking" as such if it is just between the 2 of you. but it provides the sharing functionality of Facebook and applies the principle to your romantic relationship only. Another example is FamilyLeaf, a Social Network for you and your family. It allows you to be part of several families networks simultaneously, such as your own, your partners, and your step father/mother's.

Dunbar's number is named after the British anthropologist Robin Dunbar, who first proposed this number in 1992. It refers to the suggested cognitive limit to the number of people with whom one can maintain a stable Social Relationship. As the number of Social Relationships increases, the quality of these relationships is diluted since the time requirement grows beyond the point where we can effectively maintain them. Obviously, in 1992, Social Media did not exist. The Internet itself was still in its infancy. So the question arises: in the world of Social Media, does Dunbar's Number still apply?

In early 2012 a study was undertaken by a blogger on the well known technology site Gizmodo.

FACEBOOK

Is Dunbar's Friend-Limiting Number Still Relevant in the Facebook Era?

He set out to disprove Dunbar's number by sending a personal, meaningful message to each one of his 2,000 Facebook friends. The experiment proved conclusively that Dunbar's number IS still relevant. Working alphabetically, the Blogger only made it through 1,000 of his friends before a switch in the Facebook layout that left him without access to an alphabetical list of his friends brought the experiment to a screeching halt. At this point, however, the results were already irrefutable. In order to attempt the experiment, the Blogger had been forced, due to the time commitment involved, to actually cancel plans with real-world friends. Of the Facebook friends that he messaged, most did not reply at all, and of the replies that he did get, most of them provided further evidence that Dunbar's number is still relevant. He received multiple replies that read, "Hi, I think you may have

sent me this by mistake", and "Sorry, but do I know you?".

His own conclusion reads: "In trying to disprove Dunbar's number, I actually proved it. I proved that even if you're aware of Dunbar's number, and even if you set aside a chunk of your life specifically to broaden your social capital, you can only maintain so many friendships. And 'so many', is fewer than 200."

This experiment could be criticized for its lack of Ecological Validity. There is no 'real life' scenario which would require you to scroll through an alphabetical list of everybody you have had contact with and send them a personal message. You could also argue that it was not an accurate test of Dunbar's number. The people he was messaging were essentially strangers, he had met them at some point in his life and was now sending them a personal message years down the line, having had no contact in the interim. A more accurate test would have been to attempt to maintain and build a social relationship with each individual on the list from the point of initial contact. A feat which would clearly be much easier today than it would have been even 15 years ago. Thanks, among other things, to Social Media.

A further study was conducted by the University of Indiana on the effect of Twitter on the number of Social Relationships that an individual Human Being can maintain. It is not uncommon for a Twitter user to have thousands of followers, and be following thousands of others. The study was conducted on the links created between 3 million Twitter users, over a period of 4 years. During that time the test group sent 380 million tweets. Dunbar's number counts only meaningful Social Relationships, therefore, it is not enough to simply follow, be followed by, or send the occasional tweet to another user. There must be an exchange of tweets, the Twitter equivalent of a conversation. The results showed that a new member of the Social Networking site would build their network of friends and followers up to a point at which they would become overwhelmed. At that point they would fall back and only continue to maintain the relationships where they have the strongest bond. And the saturation point? As Dunbar predicted, it lies between 100 and 200 people.

Another service provided by Social Networks, that has only recently been introduced is the ability to share your physical location by means of "checking in" to places and locations.

Existing services have added this feature, as well as Social Networking sites being built entirely on this concept. Foursquare and Gowalla are the most popular examples of such networks. While older generations refrain from using such features, it is growing in popularity among younger, less privacy conscious, generations. Such features immediately raise concerns, especially in the minds of parents concerned about the welfare of their children. While the features provided by Social Media sites have many varied uses, and are increasingly becoming an integral part of our lives, any tool that has a useful purpose can easily be exploited and misused by anyone wishing to do so. A hammer, for instance, is the perfect tool for nailing a picture to a wall. However, in the wrong hands, it can easily be utilised as an offensive weapon. Social Media, while it is the perfect tool for communication and media-sharing between friends, can also be utilised for hate-campaigns, cyber-bullying and stalking. All of which are equally as traumatic when they occur online as they are in the schoolyard, home or workplace. Location-based services, unfortunately, are one such tool that provokes misuse, and can be a stalker's best friend. When these kind of incidents begin to occur the first step is to report abuse. All of the major Social Networks make this relatively easy. It is also easy on most networks to block offenders.

According to recent studies, social media can also occasionally end friendships. According to the study, 15% of adults and 22% of teenagers have ended a friendship due to incidents that have occurred on Social Networking sites. In fact, 3% of adults have even been involved in physical violence due to incidents that occurred on Facebook. For teenagers that number rises to 8%. Physical violence may seem a excessive when the cause is merely a comment posted on a website, however, these statistics reflect the nature of the Internet as a dirty mirror of our society. The level of anonymity provided by communicating from behind a keyboard and monitor, rather than face to face, can bring out the worst in people. We say things that would never be said out loud. Sexism, Racism, Prejudice and Malice rise to the surface when there is no fear of reprisals. The worst offenders are described as "trolls". The act of "trolling" however is being tolerated less and less as the Internet becomes more entwined with our daily lives. There have been a number of high profile cases where "trolls" leaving despicable comments on memorial pages, have been awarded custodial sentences. In fact the UK is currently in the process of pushing through a High Court ruling, which would force Social Networking sites to reveal the identity of "trolls". This should serve as a warning that people may no longer hide behind their keyboards and say whatever vile thought tumbles through their head without consequence.

The threat of trolling, cyber-bullying and even physical violence pales in comparison to some of the atrocities that have occurred in the name of Social Media. Religious upheaval, rioting....... and even murder......

On a calm, peaceful evening in september 2011, piercing screams shattered the tranquility of a quiet suburban street in Surrey, England, as 28 year old Muhammad Niazi repeatedly struck his 37 year old girlfriend on the head with a hammer, in a vicious, frenzied attack. He then left her to die in a pool of blood, yards away from their petrified infant daughter. The reason? According to Niazi, his girlfriend "deserved to die", since she had been in touch with her ex-boyfriend via Facebook.

Early in 2012, a South Yorkshire woman was stabbed in the neck, a matter of hours after reuniting with an ex-lover through Facebook.

In December of 2011, Jason Redgrave, 25, shook to death the 11 month old son of Karly Hopson, a woman he had recently met on Facebook.

On June 1st 2012 Travis Brandon Baumgartner wrote on Facebook,

"I wonder if I'd make the 6 o'clock news if I just started popping people off",
Baumgartner, 21, is now wanted on three-counts of first-degree murder and one count of attempted murder, after opening fire on his colleagues during an early morning robbery.

Also in early June 2011, 15 year old gang member Ricky Moreno used his Facebook page to "call out" other gangs in attempt to provoke a fight. Later that evening a confrontation resulted in the shooting, and subsequent death, of 16 year old Nestor Alvarado.

There is one common thread that ties all of these chilling incidents together. Perhaps the perpetrators of these crimes were simply psychopaths who needed nothing more than an excuse to carry out their brutal murders. Or perhaps Social Media is stretching the boundaries of our Social Relationships too far... even beyond breaking point.

professional life

"The world judge of men by their ability in their professions, and we judge of ourselves by the same test; for it is on that on which our success in life depends"
- William Hazlitt

While you may be of the opinion that since you do not use Facebook in the workplace, Social Media does not affect your professional life at all, you may very well be mistaken.

Social Psychology is defined as: "the scientific study of the ways that people's behaviour and mental processes are shaped by the real, or imagined presence of others." Social Psychologists emphasise the core observation that human behaviour is a function of both the person and the situation. Accordingly, each and every individual contributes and expresses a set of distinctive personal attributes to a given situation. However, each specific situation also brings a unique set of forces to bear on an individual, compelling him or her to act in different ways in different situations.

When you examine your own professional life you will find that you act very differently in the workplace, to the way that you act in your personal life. Furthermore, you will act differently when you visit your in-laws, child's school or any number of situations that you may find yourself in. Almost to the point of having multiple personalities which manifest themselves in different situations. The personality that we present to our employers are polished versions of the real thing. For most people, the idea of our boss seeing the way that we act when intoxicated, on a night out with friends, is mortifying. However, it is not as easy as it once was, to hide the less attractive aspects of our personality.

In the digital age, our entire lives are recorded online for the world to see. Including our employers. The image that we present to them of a teetotal, church-going, model citizen, could be shattered in seconds with a glance at our digital profile. But how disastrous could that really be? We are all aware that, for the most part, we don't get the whole picture of our colleagues. Over time, we will get used to seeing the rest represented digitally. Should we be concerned? After all, our employers have their own personal lives as well. For the most part, the answer is no. However, some care does need to be taken. There have been a number of reported cases where individuals have posted negative comments about their job and/or employer on Social Networking sites, resulting in their dismissal. Note that in many cases, the comment was not even posted during work hours. So take heed, your employer can, and probably does, check up on you via Social Media from time to time. In the age of the Internet, you no longer have a private life. Again, the reaction to these types of comments posted on Social Networking sites is likely to decrease in severity as we become accustomed to seeing the details of each others lives 24 hours a day. Our employers will begin to remember that they are not employing soulless robots, but people, who have thoughts and feelings, and the occasional bad day.

Some employers, however, have taken snooping on their employees private lives to the extreme, by demanding the login credentials to their employees Facebook accounts for the questionable purpose of performing a "background check". The problem has reached such heights of severity that it has gained the attention of the ACLU (American Civil Liberties Union). Prompting a spokesperson to publicly condemn the practice,

"The demand for Facebook login information is not only a gross breach of privacy, it raises significant legal concerns under the Federal Stored Communications Act and State Law, which protect privacy rights and extend protections to electronic communications. As many of us begin to rely on sites like Facebook to stay connected to our friends and family, it's important for employers to keep in mind that, for most users, Facebook is a medium of private communications."

It is important to note that this is not the same as employers checking publicly posted pictures and statements on an individual profile, for most users, this content can be viewed by anyone who looks at their profile. It seems more likely that this is a case of the employer browsing personal mail. In fact, in a separate statement, an ACLU attorney compared the practice to an employer opening the postal mail of each of its employees at their home every morning. And stated that it was equally out of bounds for an employer to go on a "fishing expedition" through their employees Social Media accounts.

The problem has even come to the attention of the Social Networking giant itself, prompting a public statement from Facebook's chief privacy officer, Erin Egan,

"If you are a Facebook user, you should never have to share your password, let anyone access your account, or do anything that might jeopardize the security of your account or violate the privacy of your friends. We have worked really hard at Facebook to give you the tools to control who sees your information. As a user, you shouldn't be forced to share your private information and communications just to get a job. And as a friend of a user, you shouldn't have to worry that your private information or communications will be revealed to someone you don't know and didn't intend to share with just because that user is looking for, or wants to keep, their job."

It is worth noting that providing your current or potential employer with your password is, in fact a violation of Facebook's Terms of Service, which you, as the end user, agree to when you first sign up to the site. The relevant excerpt is taken from the Statement of Rights and Responsibilities:

"You will not share your password (or, in the case of developers, your secret key), let anyone else access your account, or do anything else that might jeopardize the security of your account."

Erin Egan then went on to state that Facebook would do whatever was in their power to fight the practice,

"Facebook takes your privacy seriously, We'll take action to protect the privacy and security of our users, whether by engaging policymakers, or, where appropriate, by initiating legal action, including by shutting down applications that abuse their privileges. While we will continue to do our part, it is important that everyone on Facebook understands they have a right to keep their password to themselves, and we will do our best to protect that right."

It has been made clear, despite this statement, that Facebook will not be attempting to sue any employers at this time. However, Senator Richard Blumenthal is drafting legislation to stop US employers demanding Facebook passwords. He believes that the law is required to stop the "unreasonable invasion of privacy". To date, Facebook has declined to make any official comment on the bill.

In May of 2012, California became the first US state to pass an independant bill banning the practice.

It is not just however, in our current jobs that Social Media can have an effect. In times past, when we attended an interview, we were making our first impression on our potential new employer. For this reason, the interview process has become very fake. This is something that everybody knows, but is very rarely commented on. When we walk into an interview, we are acting. We play a role, we pretend to be the person the interviewer is looking for, regardless of who we really are. The interviewer asks *ridiculous* "interview questions", such as, "What would you describe as your biggest weakness?"

To which we give equally ridiculous responses, such as,

"I am TOO driven, when I start something, I can't stop until it is finished".

When you step back and look at an average interview scenario, it is so ridiculous it is laughable. For most people the real answer to that question would be,

"I just can't seem to drag my lazy arse out of bed in the morning",

or,

"a big greasy kebab is my achilles heel".

However, now that our entire lives are recorded online for the world to see, the interview dynamic is changing. Potential employers can, and do, check up on us before the interview. They know all of the gory details of our personal lives before we even walk through the door. The walls have been torn down, and the facade drops. The result *should* be a positive one, unfortunately however, for the most part, it's not.

The problem is that our online profiles are not usually a well-rounded, accurate reflection of our whole personality and day to day activities. For instance, I have several jobs and work roughly 13 hours a day, 6 days a week. However I don't post endless status updates about the tasks that I am undertaking, neither do I post pictures of myself sitting at computers or serving customers. The photo's that do surface of me on Social Networking sites tend to be the more interesting ones. The ones in which I am intoxicated on nights out with friends or at rock concerts. To a potential employer, this would present a distorted and disproportionate portrayal of my life and character. Any conclusions that a potential employer would draw from browsing my Facebook profile are likely to be negative ones, at least from a business and employability perspective. The knockon effect being that my chances of securing the hypothetical job are damaged. As unfair as it may appear, it is not uncommon. As much as 37% of employers admit to viewing potential employees Social Networking profiles before making a decision on hiring. The actual number may well be much higher than that, but many may not be willing to admit it.

Social Media is also making drastic changes to the ways in which we find employment. Currently there are 2 main roads to employment. The first involves searching for employment ourselves. The other is headhunting. I am not referring to the ancient practice of taking and preserving the heads of enemies after killing them, but rather, "Executive Search" as well as traditional recruitment. In times past, individuals would provide recruitment agencies with their CV, companies would provide them with positions to fill and the recruitment agency would match the most suitable candidates to the correct jobs. It is now commonplace however, for recruitment agencies to look beyond their pool of applicants, using Social Networks to seek out suitable candidates, before using the same Social Networks to approach these individuals in an attempt to coax them from their current jobs. This approach has been so successful that recruitment agencies have been set up based entirely on this principle and "Social Recruitment" seminars are now commonplace.

The Social Network of choice for Professionals, Job-Seekers and Recruitment Agencies alike, is LinkedIn. LinkedIn actually describes itself as "the world's largest professional network". And with over 150 million+ members, it is teeming with potential candidates for any position that needs to be filled, and unlike Facebook, it presents a clear overview of their career, expertise and experience.

For this reason LinkedIn has evolved to streamline this process, adding the ability to promote job openings, and apply for them directly from the site. According to about.com LinkedIn has members from all 500 of the Fortune 500 companies. Plus LinkedIn members comprise 130 different industries, and include well over 100,000 recruiters. Recruiters themselves, however, may soon find their existence in jeopardy. LinkedIn is becoming a powerhouse for recruitment, and its strength lies in its ability to connect the employer directly with the potential employee. The days of the middle man could well be numbered.

Obviously the negative impact of this trend is the risk of excluding the potential candidates who do not have access to these facilities.

---

[5] Image Source: mariosundar.com

business

"good businesses are personal. The best businesses are very personal."

- Mark Cuban

Social Media FOR Business

Social Media Marketing is the current "buzzword". Every online marketing Guru worth his salt is pushing this. While it is often a good idea to refrain from being swept up in the hype, sometimes what is being referred to as "the next big thing", actually is. This is one of those rare occasions. Social Media, like everything else, is revolutionising Marketing online. Any business that is not currently involved will soon begin to feel the negative impact that it is having. That is if, of course, they are not already. If billions of people across the globe are involved in Social Media, it is ludicrous to assume that it is pointless to Market to such a vast audience. And even if your business is a local one, if even 5% of your potential customers are involved in Social Media, it is surely worth the effort to reach out to them.

However if you live in the UK, at the beginning of 2012 the population stood at 62.3 million, and there were 30.5 million Facebook users. That is 48.9% of the population that is using Facebook. If you consider that around 19 million UK residents are under 16, and that most business will not be targeting this age-group, that makes around 70% of the target audience of most businesses Facebook users. That is without even involving any other Social Networks.

When you look at those statistics, it becomes clear that business owners who are stubbornly refusing to involve themselves with Social Media are shooting themselves in the foot. If you refuse to move with the times, you are an ostrich burying your head in the sand, kidding yourself that the world is not changing around you.

I run a Digital Marketing company[6]. We do Web-Design as well as Social Media Marketing and SEO (Search Engine Optimization). One of our first clients was a Jacquard Scarf manufacturer who originally hired us to build a website for them. In our initial proposal, we stated that, included in the price, we would manage a Facebook page for them for 6 months. It has since become clear that that proposal was not thoroughly read. After launching the Website, we arranged a meeting to present to our client, the Facebook page we had set-up. Much to our surprise, before we had chance to open our mouths, the CEO dismissed the idea of a Facebook page, claiming that they didn't want it. To quote the owner "Facebook is something my 12 year old niece uses all the time. This company doesn't want to be associated with it". This attitude appears counterproductive, surely in business, as in nature, adaptation is key to survival? Does anybody *really* think that the internet is just a fad? Or that people are going to stop using mobile phones? The idea is ludicrous! It's pure arrogance.

If you look at some high profile companies that are struggling, it's easy to see where they could have been saved by embracing change instead of ignoring it. Had the Royal Mail moved with the times, we would all be using @royalmail.com email addresses instead of hotmail and gmail. We could all be streaming our movies direct from Blockbuster instead of Netflix or LoveFilm. The list goes on. And these are massive companies, who spend millions on analysing trends in the market. Which begs the question: How did they miss out?! I appreciate that Blockbuster is now offering movie rentals

---

[6] The Xanadu Marketing Group. www.xanadumarketinggroup.com

online, but it's too little, too late.

I understand of course, that they may have been blindsided by the speed of change. The commercial internet is less than 20 years old, and it has revolutionised every aspect of our lives. There's no going back now. And now, with the Web 2.0, Social Media is revolutionising it all over again. If I were to open a bricks-and-mortar grocery store tomorrow, how long do you think it would take to challenge the chain stores like Morrisons and Wal-mart. Or even just to become a blip on their radar? When Google launched, it overtook the market leader, Alta-Vista, in less than 3 years. Facebook, at 8 years old, has just had an IPO at a valuation of $100 billion. I read a post online recently, that compared the rise of Social Media as a marketing platform, to the rise of AOL. It was stated that advertisers flocked to AOL because that's where the people were, before quickly abandoning it in droves. The author proceeded to pose the question: "Wouldn't you rather drive your own car down the information superhighway, rather than hitch a ride on the back of somebody else's bus?" That logic appears to be fundamentally flawed. The question assumes that the Marketing tactics of running your own website and marketing via Social Media, are mutually exclusive. We either run our own website/blog, OR we run a social media campaign. However, every single article ever written, and instructional video ever made, on the subject of Social Media Marketing, makes the assumption that you are doing both simultaneously. They feed off of each other. That's why social plug-ins exist for you to embed into your site or blog. So the obvious conclusion to draw is that Social Media is less like catching a ride down the Information Superhighway on someones bus, but more like putting up billboards along it.

Incredibly versatile, highly interactive and engaging billboards. This should serve as a wake-up call to business owners everywhere. The world is changing, you must embrace it now or suffer the same fate as Woolworths, Borders, and the entire newspaper industry. Consider yourselves warned.

The term "Social Media Marketing" is a little misleading however. The presumption is that the goal is to promote your product or service. An increase in sales, however, should be a by-product of your Social Media efforts. If you make it your primary concern, it would be a little like going to a party, and sitting down every other guest individually and pitching your new business idea to them. This is not socially acceptable behaviour and would not win you any friends, much less investors or business partners. The goal of Social Media, as a "Marketing Platform" for business, is exactly the same as the goal of Social Media for the individual, and that is communication. Social Media should serve to humanise large corporations, as well as small businesses, making them much more approachable and engaging. It's more about customer service than promotion, but the by products of this will be an increase in brand awareness, and as such, sales.

Social Media AS a Business

So we have examined the positive impact of Social Media on a wide range of businesses, but how does Social Media fare as a business in and of itself? Facebook went public amid a media storm on 31st May 2012, at a valuation of $100 *Billion*. That, to date, is the largest IPO in tech history. That fact, on its own, presents a very positive outlook on Social Media as a business. But it is not all good news.

Since its IPO Facebook has been subject to severe scrutiny resulting in a backlash from the overwhelming majority of its new investors. Prior to the public offering, Facebook and its bankers made the decision to raise the size of the offering by 25%, forcing early stage investors to surrender more of their shares. They simultaneously raised the price. Demand was so great that at the critical moment, when Facebook was due to begin trading on public markets, the software crashed, leaving many investors unable to purchase stock. Immediately following the IPO, the stock began to decline in value. According to the wall street journal the debacle was a case of "if you can't get it, you want it, and if you *can* get it, you definitely *don't* want it." The sheer size of the IPO created too much supply, but not enough demand, and when that happens, there is only one place for the stock price to go, and that is down. In the wake of the IPO debacle, shareholders have filed a class action lawsuit against the Social Networking giant in an attempt to recover their money. They claim that Facebook's prospectus contained false information.

Will Facebook recover?

It is worth noting that the best investments are determined, not over a matter of hours and days, but over months and years. While Facebook's stock price dropped after its Initial Public Offering, the real test is to see where the price goes over the coming months and years.

Facebook exists in 2, mutually-inclusive forms. Facebook as a service, and Facebook as a business. These 2 forms are co-dependent, one cannot exist without the other. Facebook as a service, cannot survive without the financial support provided by Facebook as a business. Just as Facebook as a business cannot generate an income without the user base catered to by Facebook as a service. If one falls, so does the other.

Facebook as a service however, is peaking. Its growth is slowing, and in some areas, its user-base is even dwindling. The long-term survival of Facebook as a service will depend on its ability to integrate seamlessly with existing products and platforms. Just as we have seen with the recent announcement of the iOS 6, the latest incarnation of Apple's popular mobile operating system, upon which all of Apple's mobile products are built. The iOS 6 has deep Facebook integration. Unusually, Apple is actually not the first to implement this type of integration. Microsoft's mobile platform, Windows Phone 7 has had deep Facebook integration since launch. Likewise, the upcoming version of Microsoft's desktop Operating System, Windows 8, also has Facebook built directly into every nook and cranny. However, the vast majority of Facebook's revenue stream is generated by advertising on its desktop site. In fact, over 85% of Facebook's income is generated by these advertisements. So, as less and less of Facebook's users log on to the website in order to interact with Facebook, its profitability will take a nosedive. This may seem counterproductive, in securing its user base for the foreseeable future, Facebook has sacrificed its revenue stream. Therefore, in order for Facebook to survive as a business, it must find a new way to monetize. We must assume that the officials in charge at Facebook must have seen this day coming long ago, and, therefore, have a contingency plan in place. One can only speculate that this is the reason behind the existence of "Facebook credits", Facebook's very own virtual currency, which must be purchased initially with legal tender. If, however, Facebook *is* nothing more than the current fad, and the king of Social Networking does eventually implode, what will this mean for Social Media as a whole?

With Facebook's financial stability uncertain, what of the other titans of the Social Networking scene? Twitter has not yet had an IPO, but it is currently valued at an estimated $10 billion, which is a staggering amount when you consider that, since its launch in 2006 Twitter has not yet made a single penny in net profit. As well as having 3 different CEOs in 3 years. It seems most likely that Twitter will eventually monetize by leveraging its huge celebrity user with promoted tweets.

The financial struggles of these companies raise far wider concerns than the longevity of the individual companies themselves however. The largest contributing factor to the "dotcom crash" of the nineties was the extensive, massive overvaluation of internet companies. Are we witnessing a repeat of history with Facebook valued at $100 billion as it is poised to lose its revenue stream, and Twitter valued at $10 billion despite never reaching profitability? Is the seemingly unjustified, astronomical valuation of these companies a warning sign that we are hurtling headlong into a second "dotcom crash"? If so, who will survive to carry Social Media into the future? With Google being one of the survivors of the last "dotcom crash", will they endure a second time, leaving Google+ to carry the mantle of the new king of Social Networking? Perhaps fresh, new contenders will rise from the ashes of their fallen predecessors to usher in a new age for Social Media.

Or maybe we will just learn to speak to each other again.

I know.......

It's a long shot.

entertainment

"This art of resting the mind and the power of dismissing from it all care and worry is probably one of the secrets of energy in our great men."

- Captain J. A. Hadfield

My favourite philosophy and personal motto is the phrase: "Work to live, don't live to work."

With that in mind, what do we do when the working day is over and we are once again in total control of our own time? The average adult spends about 4 and a ½ hours every day watching Television, as well as 2 and a ½ hours surfing the net. According to the Telegraph, more money has been spent of Video Games in the UK over the last few years, than on films, including both trips to the cinema and the purchase of DVD/ BluRay. This being a clear sign that gaming has grown beyond the target audience of children and teenagers, and is now an activity shared by the entire family. With the Economic downturn, shopping is not as popular as it once was, but surveys suggest that it is an activity in which we still love to partake when we have disposable income.

So how is Social Media affecting our entertainment? Much of the entertainment that we enjoy is moving online and will continue to do so. More and more devices are being connected to the internet in an inherently social way. Let's examine the top 3 relaxation activities in more detail.

Currently, of the 3, Television is affected by Social Media the least. But that does not necessarily mean, not at all. Moreover, as time passes, our consumption of TV will become incrementally more social, to the point where it becomes one of the *most* affected by Social Media. This is a paradox since the concept of TV as a social experience is contradictory to its very nature. Cinematic consumption is an inherently anti-social activity. We shut ourselves in darkened rooms watching movies and television shows with minimal social interaction, if indeed there is any at all. In fact during a visit to the cinema it is considered rude to converse at all and you are even required to switch off your mobile phone for the duration of the film. A breach of these rules could result in expulsion from the theater. This is designed to create a more immersive experience for the moviegoer. However, film and television can provide an inexhaustible supply of conversation topics *after* the closing credits have rolled. And of course anything that is worth discussing, as well as many topics that are not, are being discussed at great length, 24 hours a day on Social Media sites. Discussion of films and Televisions shows is so popular that literally thousands of forums and interactive blogs exist, dedicated to the subject. Some covering the subject as a whole, others focussing on specific genres and some even focussed entirely on a single film or show. This lengthy discussion results in the phenomena known as "Social Discovery".

Social Discovery describes the process of individuals or groups learning of the existence of something (anything from films to fashion) through its exposure on Social Media platforms. More and more, our Social Graph is influencing our choices of entertainment. This is turning linear TV guides into a thing of the past as our friends, family and Social Graph help us decide what to watch. This effect will become more prominent as more of the content itself moves online and services offering TV and film content such as Netflix and Hulu having deep social features, as well as the inbuilt ability to share our preferences on other Social Networks such as Facebook and Twitter.

Not only is our TV content going to our Social Networks, our

Social Networks are coming to our TV sets. With content moving online, our TV's require native network connectivity. Many TV's are hitting the market with this feature, as well as huge numbers of set-top boxes providing this additional functionality. TVs having internet access has paved the way for the introduction of Apps to our TV screens. Many of the few Apps that are currently available on these platforms are, in fact, Social.

Some film-studios want to go a step further though, by harnessing the power of Social Media they want to flip the anti-social nature of cinematic viewing on its head. In 2011 tech giants Intel and Toshiba partnered to sponsor the first ever social film, The Inside Experience.

The Inside Experience is a horror movie that follows the story of Christina. Christina Perasso is a tough, resilient, 24-year old girl. She wakes up trapped in a room and has no idea where she is being held or who did this to her. She has been left access to her laptop. Through the use of an intermittent wifi signal, she reaches out to friends, family and the general, growing audience through social media. She lists facts, clues, pictures and videos to aid the audience in figuring out where she is, who her captor is, and why she was kidnapped.

Directed by D. J. Caruso (Disturbia, I am Number Four) and starring Emmy Rossum (The Day After Tomorrow, The Phantom of the Opera), The Inside Experience was broken up into segments. After the release of the first installment the general public was able to contact Christina, as well as members of her friends and family via Social Media sites including Twitter, Facebook and YouTube, all in real time, attempting to help Christina discover her location as well as uncover the identity of her captor. Christina would post pictures and short videos to her Social Networks to aid the public. The audience participation affected the direction of the film as each installment was released. Some of the videos sent back to Christina were even selected for inclusion in the final film. Upon discovery of Christina's location, many local participants even went to watch her "escape" live, some of these fans were taken by "FBI Agents" and "interrogated", more footage which ended up in the final film.

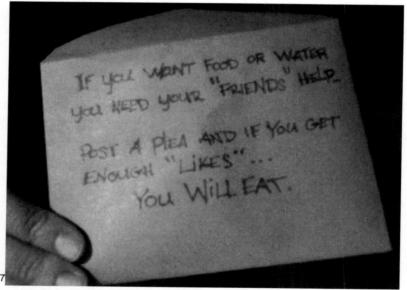

My combined personal love of both technology and horror left me no choice but to take part. While the experience was both revolutionary and thoroughly enjoyable, labelling it "the future of cinema" was absurd. As much as I enjoyed taking part in what is essentially a piece of history, the idea

---

[7] One of the pictures posted on Facebook by Christina

is completely impractical for mainstream audiences. The film unfolded in real time over a period of 11 days, in comparison to the traditional runtime of around 90 minutes. The first installment was unveiled on the 25th July 2011, kick starting the experiment. From that moment, my life ground to a halt. It did not start again for 11 days. To be completely immersed, or even just to stay up to date, would require an individual to not leave his or her computer for the entire period. Even then, it is difficult to monitor several Social Networks at once. The only time I was not involved was during brief, but necessary, periods of sleep. Unfortunately however, there were millions of other participants from around the globe. While I was sleeping, they were not. I would wake up everyday to find I had some serious catching up to do. Most people however, cannot simply drop 11 days from their lives whenever it takes their fancy, plus, taking part in an experiment to make film more social, actually had the result of making me antisocial toward my real-life friends and family. I would not go as far as to suggest that I would never be a participant in similar forms of entertainment in the future, but it will never replace a film that you can just sit in front of for 90 minutes of mindless enjoyment.

As previously observed, a wider range of people are beginning to take pleasure in Video Games. The rise of gaming as a family activity has inspired gaming consoles, designed specifically to be more social. Examples of this are the Nintendo Wii, and Xbox Kinect (an add on rather than a standalone console). The result is a transformation of the stereotypical view of gaming as a hobby, from this:

(Image Source: www.videojug.com)

to this:

(Image Source: www.familyvaluesclub.com)

Almost every gaming console on the market has networking capabilities, if not "out of the box" then through a separate piece of hardware that can be obtained at an additional cost.

The main purpose of this is to make it possible to play with, or against, other people from around the world. Most also have chat functionality, via instant messaging, VOIP (Voice Over IP) or both. As gaming is becoming more social by the day, it can be greatly complemented through the use of Social Media. Smartphone and Tablet devices are now viewed seriously as gaming platforms and over 60% of App downloads are games. The vast majority of these games, like their console counterparts, have a social layer, allowing the user to play with or against their friends. And what is the platform of choice that is being utilized to connect players? It is, of course, Facebook. I am currently working on an update to my own iPhone game, pShooter, adding Facebook connect to enable users to play against their friends.

Facebook itself is growing in popularity as a gaming platform, with many third party companies producing games solely for Facebook. According to allfacebook.com 53% of Facebook's 900 million users play games on Facebook and 19% even claim to be addicted. With half of Facebook logins specifically for the purpose of gaming and 20% of players parting with hard-earned cash in exchange for in-game benefits, developing games for Facebook can be extremely lucrative. Proof of this can be found in the form of Zynga, the game development company behind the absurdly popular games, CityVille and FarmVille, which are the 1st and 2nd most popular Facebook games respectively. CityVille alone has over 40 million monthly active users. Zynga filed for a $1 billion IPO in July 2011. The total revenue for the quarter before it went public was $311.2 million. After existing for only 4 years, the company has surpassed the market value of EA Games.

Shopping is an activity that can be enjoyed in 2 ways. Either through a physical visit to a shopping center or mall, or online, from the comfort of your favourite chair. You would not expect Social Media to have a particularly large effect on a physical shopping trip. There are however, many retailers, taking to Social Media in an attempt to lure customers away from their computers and back onto the high street. Many offer discounts, bargains and voucher downloads to customers who "like" their Facebook page or "follow" them on Twitter. They are also slowly beginning to leverage Location Based Services such as Foursquare, providing them with exposure every time a customer "checks in" to their outlet.

8

By offering in-store incentives to consumers who connect with them via Social Media, they extend their network and more importantly, bring customers back into their stores.

---

[8] Example of a Foursquare checkin sticker for retail outlets.

The benefits of choosing the alternative and doing your shopping online are many. The largest of which being the far wider range of products to choose from, as well as the ease with which price comparisons can be made between multiple retailers. And littered across almost every e-commerce page you will ever see are invitations to share the products on a wide variety of Social Networks. Recent surveys have found that over a quarter (27%) of online shoppers under the age of 35 share their purchases on Social Media sites. Furthermore, 54% of the same age group claim to rely on shopping tips provided to them by friends on their Social Networks. Up to 65% of the entire group admit that they are more likely to make a purchase if they receive a recommendation via Social Media. With Social Discovery being such a powerful force driving of sales for online retailers, you can expect them to place even more emphasis on their Social Media campaigns in years to come, so we can expect to see this trend continue indefinitely.

relationships

"Love comforteth like sunshine after rain."
- William Shakespeare

Over the course of human history, the way in which romantic relationships have developed has also been constantly changing and evolving, again, driven in the largest part by advances in technology. In recent years, the early stages of relationships have often been conducted or complemented with text messaging. This trend has continued to evolve with the rise of Social Media, and it is now commonplace for this early stage of the relationship to be conducted via Social Networking sites, primarily Facebook.

This trend needs to be put into perspective. During the 90s, chat rooms were demonised by parents. This was due to a small number of high profile cases where teenagers were kidnapped by adults who had been posing as teenagers. Let's be clear, this is NOT the same thing. The anonymity of chat rooms that allowed for this level of deception is not reflected on the Facebook platform. Fake profiles are extremely uncommon and the full profiles and highly visible list of mutual friends make them instantly recognisable for what they are.

The term "Facebook Official" is common among young users of the Social Network. Its popularity has reached a level where major events are posted instantly, and it is a woeful, but widely held, belief that a relationship is non-existent unless it is reflected in the "Relationship Status" of a persons Facebook profile. It has become the modern standard for making such announcements.

Giles Lloyd

🚂 Studied at king edward 6th
🏛 Lives in Leicester, United Kingdom
9 ♥ In a relationship

Social Media has also changed the dynamic of established relationships, albeit to a lesser extent. Social Networks have even been created solely for couples to privately share content with each other. Cupple for the iPhone is one such example.

---

9 Look at me..... I'm Facebook official!

Cupple is like Facebook, except the only person in your friends list, is your partner. You share everything with them, creating an intimate relationship online to reflect that which you share in real life.

There is a stigma attached to asking a member of the opposite sex for their phone number. In most scenarios, if there had been no prior contact, this is a suggestion of attraction. For this reason most individuals would become upset to find that their partner had given their mobile number to a stranger, at a social function or on a night out. Adding someone on a Social Network however does not carry the same stigma. Although the motivation for adding a stranger on Facebook after meeting them on a night out, in many cases, is likely to be the same. But since it doesn't carry the same stigma, many people feel that they have no right to be upset with their partner for it. This type of activity has led to a rise in paranoia, and can upset the balance of a healthy relationship.

A trend that is becoming popular among teenagers, to combat

this type of paranoia, is password sharing. When they become involved in a monogamous relationship, they swap the passwords for their email accounts and Social Networks with their partner. It is a symbol of mutual trust and faithfulness. Providing their partner with access to all of their private messages and emails shows that they have nothing to hide. This trend is often condemned by older generations, who value their privacy far more than their younger counterparts.

Some of the arguments against password sharing include:

- Paranoid partners could become obsessed, spending hours searching through emails and private messages, looking for the clue that could prove their partner's infidelity,
- If the relationship doesn't last, less mature individuals may misuse the access they have to their ex partner's accounts to send malicious messages or post malicious content.

Both of these arguments are weak at best. Surely the act of providing access to private accounts is proof enough that an individual has nothing to hide. For this reason, the partner would not need to invade their privacy and look at their private messages at all? Furthermore, if an extremely paranoid person *did* decide to spend hours scouring their partners private messages for proof of unfaithfulness, and found nothing, would that not put their mind at rest?

While I accept that many individuals, especially the younger generations, in which this trend is common, can be incredibly immature, especially after a breakup, there may well be a simple solution to the second argument. If you end a relationship with someone who has access to your accounts.......... change the passwords. I suspect that doing so may stop jealous ex lovers from misusing their access. It may sound crazy, but try it, I think it will work.

the future

"The habit of looking to the future and thinking that the whole meaning of the present lies in what it will bring forth is a pernicious one. There can be no value in the whole unless there is value in the parts."
- Bertrand Russell

As far as we have come, we have still barely scratched the surface of what technology can offer us, and the ways in which it can change and enrich our lives. And the same thing can be said for Social Media, its meteoric rise has just begun. Try to imagine how we will be living 100 years from now. And think about the ways in which we will be communicating and sharing our lives. If Social Media has revolutionised those things in such a dramatic way in a few short years, imagine where we will be in the next century.

It is easy to predict the changes that are coming in the not-too-distant future however, the trends are already taking us there. Everything is moving online, from shopping to booking holidays, and this trend is likely to continue. Starting with television.

The slow migration of television into the online world began a few years ago, with the introduction of services like the BBC iPlayer. Other TV networks followed suit with 4oD and Demand5. While most of these services will still be accessed over the Television networks, they are all also available online. It is now easy to stream these services to your desktop or mobile phone. Set-top boxes have also been available for a long time that allow you to connect your television to your home WiFi network and stream these services from the Internet to your Television set. You can even achieve this through games consoles such as an Xbox, Playstation 3 or Nintendo Wii.

Recently we have seen a big push from all angles, setting the stage to take Internet television to the next level. Firstly, major networks have started producing shows to be aired online only. You cannot watch these shows through regular Television or Cable networks. The big players in technology have also begun making set-top boxes that take streaming to a new level. Apple have created Apple TV, and even Google have created a set-top box of their own. These products allow you to stream media to your television both from within your local network, and from the wider internet. They have also introduced apps, similar to the apps you can download on your smartphone. The apps on these set-top boxes add support for YouTube, allowing you to browse all of YouTube's content on your Television, and also for Facebook and other Social Networks, allowing you to share what you are watching with the click of a button. Another big player in this area is D-Link. Their set-top box is called Boxee and, like Apple and Google TV, it supports streaming from your local network and the Internet. Unlike its competitors however, Boxee has social features built into the device itself. When you register your Boxee and connect it to the Internet, you become part of a global social network of Boxee users, should you decide to utilise these features, you can follow the activity of other Boxee users, rate and recommend shows or Movies and get recommendations in return. Boxee will also pull videos from links in the activity feeds of your existing Social Networks such as Facebook and Twitter, it will add these to your watch queue to be watched from your television set at your leisure.

Set-top boxes and games consoles are no longer, however, the only way to get Internet content onto our televisions. Many high profile Television manufacturers are beginning to build these features directly into their Television Sets. This allows you to connect your Television to your local network as well as the internet, without the need for third party hardware.

The final push for moving television online is coming from the increasing number of services that are springing up online, from which you can get your content. We have already mentioned existing Television stations that have put their content online as an addition to the services that they already offer. There are however many services which only exist online, for the sole purpose of streaming content to your devices. Some well known examples are Netflix, Hulu and Amazon Instant Video.

Netflix, which is primarily for streaming movies, has been

---

[10] The Boxee interface. Note the "Friends Activity" option.

growing in popularity in the US for a number of years, and is just beginning to get a push into the UK market with major marketing campaigns on UK TV and radio stations. It is somewhat ironic that a service which is designed to kill off traditional television is using television advertising to fuel its growth. Netflix also has deep Facebook integration for sharing what you are watching and finding new content for you to watch based on recommendations from your friends.

Hulu is an online service for streaming television shows. While it has thousands of Television shows from many major networks, it also has begun to create some original programming. Currently Hulu has no Social Features, although it does provide its users with personalised recommendations based on shows that they have previously watched. Currently Hulu is only available within the US and Japan. They are currently working on the legal issues which will allow them to expand internationally.

Amazon Instant Video has a roughly even split of television shows and Movies for you to stream to your devices or set-top box. New episodes of current television programs are available on amazon instant video the day after they air. Amazon Instant Video gives you the option to Rent or Buy the latest movies, allowing you to stream the movie once for rentals, or as many times as you like. You can also download bought movies to your computer's hard drive. Like Hulu, Amazon Instant Video has no integrated Social Features.

With the large push Internet Television is receiving, it will only be a short few years before this becomes the primary way that we watch television. And, whether the integration is provided by the service itself, or the devices we use to consume the media, it is clear that Social Media will be at the heart of the experience. Allowing us to share our preferences with our friends, rate, and discuss, movies and television shows, and recommend something new.

The second change that Social Media is likely to bring in the coming years is to telephones. When you give it some thought, the idea of using 11 digit telephone numbers seems archaic. It is likely that this will be abolished as, like television, telephone services are moved online.

There are already many services that exist, that will allow you to make a telephone call from an internet connected device, whether that is a computer, tablet, smartphone or even a television. The most well known of these services is Skype.

The Skype software allows you to make a call to other Skype users via their username. You can also make calls to traditional phone numbers. The recent Skype - Facebook partnership allows you to make calls to anyone in your Facebook contact list, as long as they also have Skype installed, via their actual name. As time goes on, this is likely to become the norm, until all calls are made to someone's actual name and the traditional phone number is all but extinct. Other major services include Google Voice.

Social Integration into smartphones is making the experience of communicating via Social Networks as seamless as traditional text messages. I have a Windows Phone 7 device, and all of the text messages, windows live messages, and Facebook chat messages from each contact appear in one conversation. It is sometimes difficult to tell which service was the source of each message. Replying to each requires the exact same process, the service you are messaging is irrelevant. Each of the major smartphone platforms have similar integrations of Social Media, making sms all-but redundant, along with the 11 digit telephone number required to make it work.

Most smartphone platforms also sync contacts from your Social Networks as well, so if you have Skype or another VoIP (Voice Over IP) app installed, all that is necessary is a data connection, no minutes or text messaging allowance is needed, and neither are phone numbers. It appears that the only reason that phone numbers still exist is to maintain support for older devices, as well as land-line telephones. As we progress over the next few years, the old mobile phones that we need to maintain support for, will be the smartphones of today, and as less and less people use landline phones at all, the need for telephone numbers, as well as voice and sms allowance from your carrier will disappear. Many people already do not have landlines phones, choosing instead to rely entirely on their, more convenient, mobile phone. This trend will only continue, and for the stragglers, who will cling to their landline until the last, internet ready landline phones already exist.

There have been many rumors over the last year or so about a "Facebook Phone". This phone is rumored to take these trends to the next level and abolish traditional Phone principals altogether. You will simply have a data allowance with your network, every call will be made over the internet, and text message sent via Facebook messaging, your contacts will be your Facebook friends list and if you want to add someone to your phonebook, you will need to "friend" them on Facebook. Obviously this means that the phone is completely useless for contacting anyone who does not have a Facebook profile, but how many of your friends don't? This is currently only a rumor, whether it turns out to be true or not is irrelevant, because this will happen one day. With or without Facebook, this is the future of communication technology.

When we use the web, there are 2 main starting points. We are either going to a specific website, for which we type the url into the address bar of our browser. Or we are searching for something, for which we use a search engine. This is the reason why the vast majority of internet users have Google, or another search engine, as their homepage. It is in fact much easier and faster to type "fa" into Google and click on the Facebook link at the top of Google's instant search results, than it is to type "http://www.facebook.com/" into the address bar. This has made Google by far the most visited website on the internet for many years. However, recently Facebook has overtaken Google to take the number 1 spot.

[11] A concept image for the Facebook phone courtesy of iDownloadBlog

Facebook wants to be your homepage, it wants to be the Sun at the center of the solar system that is the internet, and have everything revolve around it. The problem is that Facebook is lacking 1 thing, the thing that tips the scale in Google's favour when millions of people are choosing their homepage. No matter what, 99 out of 100 visits to the internet will, at some point, require a search engine. For this reason it is speculated that it will not be long before Facebook launches one. Imagine the impact that this will have on the searches that you make online. Taking the internet out of the equation, if you are looking for a product or service in real life, you may use the yellow pages, or a newspaper, but more often than not, you will use the ones recommended to you by your friends or family. Facebook has your entire social graph to leverage. It can filter and sort your search results based on similar searches made by everyone you know, as well as relevant websites that they have visited and "liked" etc.

Even if Facebook does not develop a search engine of their own, Google is still by far the most popular search engine in use today, and they have recently announced Google Search+. Google Search+ is Google's attempt at doing the same thing, leveraging the social graph of its own Social Network, Google+, and filtering your search results accordingly. So it seems that whatever happens, before long even your searches will be inherently social.

afterthought

"Second thoughts are ever wiser."
- Euripides

As Social Media becomes increasingly important to us in our day to day lives and activities, what does it mean for us as Human Beings?

Humanity has a tendency to push too far, both collectively and as individuals. We need our cars to go faster and faster, resulting in millions of deaths every year due to Road Traffic Collisions. We need our bodies to perform better than ever before, hence the introduction of anabolic steroids. We push our finances further than they should go, resulting in a global financial crisis.

But is it possible to push our need for Social Interaction too far? It was not enough to be able to converse with others face to face, so we invented the telephone. Not being satisfied with the ability to telephone people from our own homes, we invented the mobile phone, enabling long-distance communication anytime, anywhere. Still not satisfied, we dreamed up text messaging, allowing communication whenever talking was inconvenient. Then, when voice and text communication was *still* not enough, we made use of the internet to share media content. And now we have reached the point where the lives of everyone we know are laid bare every time we reach into our pocket to check the time. This is known as "hyperconnectivity". Being hyper connected simply refers to the fact that we are now always connected to everyone and everything via the internet and phone-networks 24 hours a day, we are suffering an information overload. Even this is not enough it seems. Google has announced a product called "Google Glasses", a set of glasses which will use augmented reality to overlay our feeds, news and messages on top of everything else we see. Taking all of this out of our pockets and off our computer screens, and putting it right in front of our eyes every single second of every single day. The question arises, what long-term effects will hyperconnectivity have on the Human Psyche?

With the insignificant thoughts and details of the lives of every other individual we know being forced into our brain at a faster rate than we can formulate our own thoughts, we begin to reach a stage where we can no longer accurately distinguish our own thoughts from the collective. At this stage we begin to lose our sense of self, and feel like we are no longer an individual, but rather part of a collective group psyche. Our thoughts, feelings and even actions, are no longer our own, but rather, the thoughts, feelings and actions of the group. This loss of individual accountability is a major contributing factor to mob-mentality, it is what drives usually civilised, law-abiding citizens to take part in extreme acts of violence and destruction without a second thought. We have already seen Social Networking sites used to incite and coordinate widespread riots throughout the UK in August 2011, as well as the Occupy Wall Street Movement and Civil Unrest in the middle east. Although, it should also be pointed out that Social Media sites played a large part in helping to identify, track down and prosecute offenders.

The UK riots saw large numbers of citizens with no previous criminal record and no reason, resort to rioting, arson and looting of unprecedented levels. Along with 5 deaths, an estimated £200 million worth of property damage was incurred.

Is this the future that Social Media is helping to shape? Rather than creating a unified world, a utopia where humanity finally sets aside their cultural and religious differences to work together, will it instead, ultimately result in the dramatic collapse of civilization into anarchy and ruin?

Probably not... But who knows?

14808295R00050

Printed in Great Britain
by Amazon.co.uk, Ltd.,
Marston Gate.